Fannie Isabel Sherrick

Star-Dust

Fannie Isabel Sherrick

Star-Dust

ISBN/EAN: 9783337038298

Printed in Europe, USA, Canada, Australia, Japan

Cover: Foto ©berggeist007 / pixelio.de

More available books at **www.hansebooks.com**

BY

FANNIE ISABEL SHERRICK.

"God's poet is silence, his song is unspoken,
  And yet so profound, so loud and so far,
It fills you, it thrills you, with measures unbroken,
  And as soft and as far and as fair as a star."
                              —*Joaquin Miller.*

CHICAGO, NEW YORK, SAN FRANCISCO:
BELFORD, CLARKE & CO.
1888.

TO MY DEAR FRIEND,

Mrs. ELLA WHEELER WILCOX,

TO WHOSE ENCOURAGEMENT MUCH OF MY SUCCESS

IN LITERATURE IS DUE,

THESE POEMS

ARE AFFECTIONATELY DEDICATED.

CONTENTS.

MISCELLANEOUS.

	PAGE.
DEDICATORY TO ELLA WHEELER WILCOX,	9
A SEA FLOWER,	80
AFTER THE RAIN,	73
AN AUGUST NOON,	58
CALIFORNIA,	28
DREAMS,	26
EASTER SONGS: The Flower Song,	42
" The Bird Song,	43
" The Earth Song,	44
" The Songs of the Bells,	45
" The Voice of the Lilies,	47
" The Christ-Song,	49
ECHOES,	51
FAILURE,	20
FIREFLIES,	39
IN THE RAIN,	71
IN THE TWILIGHT,	23
JUNE,	55

CONTENTS — Continued.

	PAGE.
LOOKING BACK,	65
MIRROR LAKE, YOSEMITE,	36
NIGHT,	25
NIGHT ON THE SIERRAS,	34
OUR KING,	89
SEPTEMBER,	60
SHE AND I,	76
SILENCE,	12
SNOW-CROWNED,	53
STAR-DUST,	11
"SUFFER AND BE STRONG,"	86
SUNSET,	70
SUTRO HEIGHTS,	31
TAMALPAIS,	16
THE AISLES OF PAIN,	74
THE BLACK CAÑON,	82
THE CRESCENT MOON,	78
THE MECHANICAL AGE,	13
THE NEW YEAR'S DREAM,	67
THE OLD YEAR,	62
TO A DEAD BIRD,	85
UPWARD,	18

SONGS OF LOVE.

A DREAM OF THE EXPOSITION,	104
AUF WIEDERSEHEN,	151
BRIDAL VEIL FALLS, YOSEMITE,	94
COMING,	172
FORGETTING,	145
GOLDEN-ROD,	149

CONTENTS — Continued.

	PAGE.
GOOD-BYE,	138
HOPE,	168
L' AMOUR,	91
LIFE,	156
LOVE'S RETROSPECT,	140
MINE,	143
MY IDEAL,	135
MY PRAYER,	169
ON THE WILLIAMETTE,	141
ORANGE BLOSSOMS,	111
PALMISTRY,	120
PANSIES,	147
PARTED,	131
RED MOUNTAIN,	166
STRANGERS YET,	123
SOPRIS PEAK,	159
THE ELECTRIC LIGHT,	164
THE GOLDEN GAIL,	133
THE LADDER OF PRAYER,	175
THE LAKE,	170
"THE MERRY WAP,"	113
THE MOUNTAIN STAR,	157
THE NEW MOON,	153
THE OLD AND THE NEW,	99
THE PAINTER'S MODEL,	117
THE RANCHMAN,	173
THE SUMMER'S NOON,	137
THE WALTZ QUADRILLE,	109
TWIN BORN,	162
TO A ROSE,	96
TWO YEARS,	128

TO ELLA WHEELER WILCOX.

I sit in the twilight and wonder, dear,
   Of you in your Northern home,
And my thoughts speed on like the ships at sea,
   In the track of the silver foam;
But I know full well that a golden chain
   Is fettered from heart to heart,
And the spirits once joined by the links of love
   Can nevermore drift apart.

A picture, dear, in the twilight here
   The sunset has woven for me;
And I sigh — for the picture fades from sight,
   As the light on a summer sea.
Your soft, brown eyes are so full of love,
   And warm as they always are,
And your soul shines through their shadows, dear,
   Like a beautiful, radiant star.

DEDICATORY.

Sweet lines of your poems come to mind,
  Like melodies heard and lost;
Or the fitful gleam of fleeting stars
  To the mariner tempest-tossed.
And I wonder if you, in your Northern home,
  Are singing those dream-songs still;
You, with your measure of smiles and tears,
  You, with your woman's will.

The glow of the sunset is in your eyes,
  Framed in the dusk of night,
With a crimson flush on your lifted face,
  Like the flame of the fading light.
Afar in the skies a star shines out —
  Do you see it, I wonder, too? —
And is it a sign that our friendship, dear,
  Will ever be tender and true?

STAR-DUST.

Through spaces infinite the circling stars
Leave dusty trail; and earth, with measured flight,
Speeds onward in the path of unseen gold.
Her gorgeous sun king, through the glittering dust,
Looks earthward, and his fiery glance at eve
Inflames the West. The cosmic ether glows
With blood-red hues that turn the brilliant stars
To palest gold.

               Through ages bountiful
Great minds have lived and left a golden trail;
And lesser planets in the dust of thought
Have circled. Eternity, God's boundless West,
Is all aflame with glories from the sun
Of truth, and thoughts that live through centuries
Celestial burn. Star-souls leave meteor-sparks
That flash the truth through all Eternity.

SILENCE.

Silence is the mantle of each star,
   Woven on the mountain heights of snow.
   Silence is the mantle of our woe,
When to men our inmost souls we bar,
Standing from their ways apart and far.
   In its wordless spell pure souls atone,
   Voiceless, at their Maker's heavenly throne,
For the thoughts that holy deeds do mar.
God is Silence, and His works of might,
   Wrought in silence now and ever-more,
      Stand within the soul's white gallery,
Mutely eloquent; and in death's night,
   Carved, we see, upon His temple door,
      Silence! Symbol of Eternity.

THE MECHANICAL AGE.

To think, to know, to do —
These are the wheels of God and man,
The wheels of might since the world began ;
And the engine of Thought, with its pulse of fire,
Throbs through the Ages, and does not tire ;
In the brains of men it worked unseen,
Though the dust of centuries lay between,
And clogged the wheels with rust and crime
And martyred blood from every clime.

To think, to know, to do —
These were the swords by tyrants feared,
That swept down crowns by Cæsars reared ;
These were the spirits that worked within
Of men ground down by oppression's sin,

And lifted the masses above the king.
Lo! through the hush of the centuries ring
The voices of Thought that awakened then
In the slumbering souls of thinking men.

  To think, to know, to do —
These were the roots of Chivalry's flower,
The golden blossom of martial power,
Born in the gloom of the dark Crusades,
Crimson-stained by reeking blades,
And worn by the princely-born of earth;
But the living thoughts that age gave birth,
Hid by the shadowing leaves of pride,
Were grander far than the flower that died.

  To think, to know, to do —
These were the tools reformers grasped
To tear down shams Religion masked,
And show to men the truth divine,
Shorn of all strange device or sign;

And these the keys of price untold,

That opened wide the doors of gold,
And let in the sunlight to sweep away
The shadowed gloom of centuries gray.

 To think, to know, to do —
These are the wheels of God and man,
The wheels of might since the world began.
And the grandest age of all the years
Is this age of work that genius rears.
The engine of Thought, with its lightning power,
Is the greatest boon of man's princely dower;
For the men who think are of royal birth,
And the men who work are the kings of earth.

TAMALPAIS.*

Purple mount to God uplifted,
    Crimson-tipped and fair,
Many suns have round thee faded,
    Leaving impress rare;
    All their color blending,
    Through the far skies sending
    Glory grand, unending.

Violet clouds around thee linger,
    Rarely tinged with gold;
Mists from sunset seas enwrap thee
    In their amber fold.
    Gorgeous glow and gleaming,
    Peace and starry dreaming,
    Glorious is thy seeming!

---

* One of the mountains that guard the Golden Gate.

Peace steals upward to thy splendor
    From the earth and sea,
Sorrow loses sorrow, looking
    Sunward unto thee.
    Oh, thy precious bringing,
    O'er the pale seas winging,
    Glory, earthward flinging.

Peace-crowned height in seas of splendor,
    Holy, facing God;
Sacred unto men thy mission,
    They of grosser clod.
    Oh, the boundless measure
    Of thy sinless pleasure,
    Meed of heaven's treasure!

UPWARD.

Upward flies the lark at morn,
    Upward!
And the song of God is borne
    Through the violet deeps,
    Where the sunlight sleeps,
        To the earth-bound soul.

Upward flies the eagle king,
    Upward!
Royal born and strong of wing,
    Past all human eyes,
    Through the purple skies;
        Sunward is the goal.
Upward rise the circling peaks,
    Upward!

Face of God, the mountain seeks,
    Braving storms of woe,
    Rain and blinding snow,
        Reaching for the light.

Upward rise the souls of might,
        Upward!
Fair of thought, of purpose white,
    God-like in their way;
    Nobler grown each day,
        Truth-ward is their flight!

Lift your hearts, oh, men of earth,
        Upward!
Be your deeds of holy birth,
    Born of truth and light,
    God-like in their might.
        Live, to do and dare!

Let your lives and life-work grow
    Upward!
Through the blinding storms of woe,
    As the eagle king,
    Brave and strong of wing,
        Sunward be your flight!

FAILURE.

The golden peaks seem fair and far
    To myriad souls ascending,
Far-reaching for the mystic heights
    Of fame and joy unending.
Youth presses on with eager face,
Age creeps with slow and weary pace,
And many falter in the race—
    To earth again descending.

Ah, many climb who do not reach
    The heights to God uplifted;
And many fall from dizzy heights,
    'Mid clouds of earth blue-rifted.
For lo! yon purple peak afar,
Above whose crest the midnight star
Shines oft, unlocks its crimson bar
    To souls alone God-gifted.

But shall men say that all do fail
    Who reach not heights so glowing,
But, faint and weary, linger oft
    By streamlets gently flowing;
Or turn aside from peaks of snow
To help some brother fallen low?
And do these souls a failure know
    By deeds so humbly sowing?

The purple peaks are for the few—
    They shall know fame undying;
The broad, free plains are for us all,
    Where peace and hope are lying.
God give us strength, if we do fail,
And backward turn with faces pale,
From dizzy heights, to seek the vale
    Without regret or sighing.

IN THE TWILIGHT.

Golden flowers sink to rest
   On their emerald pillows;
Golden dreams from seas of sleep
   Float on earth-wide billows;
Earth hath sound of unseen things,
Music sweet and fluttering wings,
      In the twilight.

Tired babes, with violet eyes,
   Nod their heads in slumber;
Gentle spirits, heaven-born,
   Guard their sacred number;
Earth hath ear attuned to God,
From the sky to blossoming sod,
      In the twilight.

Purple earth gives up her gems,
    All the flower-cups filling;
Hearts of blossoms, newly born,
    With her blood are thrilling;
Earth to earth and sky to sky,
God's sweet peace is passing by
    In the twilight.

NIGHT.

Night is a garden, and each star a flower
Born of God's great love, a peace-crowned power.

Every blossom with its petals white,
Opens, filled with heaven's eternal light.

Every petal, with its dust of Love,
Shines immortal in the skies above.

Circles, wheels within the wheels of light,
Turn within the garden of the night.

Dusky veils fall o'er the shadowed ways,
Hidden from the star flowers falling rays.

Night, immortal Night, is blossomed o'er,
With the blooms of God forever more.

STAR DUST.

DREAMS.

"Dreams are but interludes which fancy makes."

In the purple dream-land lying,
    White-winged dreams
  Sleep with folded pinions fair
  In the hearts of violets rare,
Where the yellow rose, low sighing,
    Slumbering seems.

Soft, gray clouds, with sleep o'erweighted,
    Far are seen;
  And each heavy-lidded star
  Drifts through dream-seas, still and far.
Mists of gold, with peace o'erfreighted,
    Lie between.

DREAMS.

Brooding wings stretch o'er the meadows,
    Purple-barred ;
  Snowy lilies, faced with gold,
  In their bosoms dreams enfold,
Where the night-wings cast their shadows,
    Golden-starred.

In the mist land dreams are lying,
    Full of peace ;
  Weary souls give up dark care
  In the dream-land far and fair.
In the hearts of roses sighing
    Sorrows cease.

CALIFORNIA.

Oh, land of song, where summer leans,
    With eyes of dreamy splendor,
To sweep the mists from off the mounts
    With kisses warm and tender,
Your woods are cool and sweet,
    With limpid rills that laugh and leap
    Adown each wind-blown, classic steep,
    And notes of birds uprising
On song-wings strong and fleet.

Oh, land of roses, rarely dyed
    By suns forever shining,
With hues that match the flame-winged clouds,
    O'er Tyrian hills declining,

CALIFORNIA.

Upon your pulsing breast,
    The hearts of blossoms bloom and hide,
    Swept by the summer's swelling tide;
    And tree-gods woo the star-maids
In nights of calmest rest.

Oh, land of gold, whose glory lies
    Upon the face, not only,
But in the flower-swept heart, fast-hid
    Through cycles long and lonely,
Beneath your mount-walls gray,
    The Midas sun left precious rays
    In æons past, primeval days,
    That human souls, outreaching,
To you might find the way.

Oh, land of love, like siren queen,
    With emerald garments trailing
And circlet set with flower-pearls
    And dews the star-blooms paling,

You pulse the heart like wine.
 Oh, sea-won bride, the ocean's might
 Flows through your veins with swiftest flight
 And from your breath, sea-sweeping,
Leaps Love, the god divine!

SUTRO HEIGHTS.

Tamalpais leans o'er thee dreamily;
   Shadows of purple clouds nod,
There, where the old ocean mightily
   Sings to the mountains of God.
Up from the East and its dawning
   Rises the gold-eyed day,
Spreading her wings like the summer
   Over the violet bay.

Flowers rise upward like spirit dreams,
   Born of the dust at thy feet;
Songs from the far wind harps heavenly
   Echo the sea-music sweet.
Oh, that the hand of a Sappho
   Here on these lawns might trace
Sonnets to make thee immortal—
   Touched by the old Greek grace.

Seaward, the Golden Gate tenderly
    Guardeth the Child-Queen State;
Sunward the noon-day slips mistily,
Laden with golden-barred freight.
There, in the West dies the sun-god,
    Shrouded in dun and gold—
Cometh the night-queen in mourning,
    Stars in each sable fold.

Dreaming, the heart reaches longingly
    Up from the wind-beaten sod,
Unto the star-flowers, blossoming,
    Pale in the garden of God.
Upward the hills and the mountains
    Reach in the solemn night;
Thrilling, the soul follows after,
    Hushed in its trackless flight.

Tamalpais leans o'er thee dreamily;
    Shadows of purple clouds nod,

There, where the old ocean mightily
  Sings to the mountains of God.
Joy, like a star, leads the morning,
  Hope, with her smile, crowns the West;
Peace folds her white wings, forever,
  Here in this Eden to rest.

NIGHT ON THE SIERRAS.

How lonely are the silent peaks
    That rise unto the sky;
How awful are the whisperings
    Of winds that never die!

Vast solitude that fills the heart
    With loneliness and fear!
God sits in awful majesty
    Upon His night-throne here.

The Truckee's crystal waters dash
    Like silver in the light;
Its maddened waves are lashed to foam,
    And all the stream is white.

The star-crowned peaks uplift their heads
    Like gods unto the sky;
The stainless snows of centuries
    Upon their faces lie.

Impassable—they guard the gates
    That hide the golden morn;
But men cut out their hearts of stone,
    And mock their giant scorn.

MIRROR LAKE, YOSEMITE.

How like a soul thou art!
   So calm and fair,
   As, through the air,
Thy mirrored forms of rocks and trees,
Scarce stirred by wind or zephyr breeze,
   Form of thyself a part.

The dizzy cliffs and domes of gray,
   Uplifted high
   Against the sky,
Look down and see in thy calm face
The impress of a God-like grace
   That cannot fade away.

Wert thou less calm and clear,
   The pictures fair,
   Reflected there,

Would never meet the human eye ;
But thou hast pictured earth and sky
  In shadowy pool-depths near.

Each human soul, like thee,
    Is but a glass,
    Where pictures pass
From day to day ; and every care
And every joy is mirrored there,
  For God and man to see.

Upon those waves of light
    The lines of thought,
    In shadows wrought,
Lie mirrored oft ; and, like to thine,
Each picture speaks the truth divine
  That marks the spirit's flight.

This morn, the orb of gold
    Crept, still and far,
    A golden star,

And on yon crested mount it hung:
By poets' eyes unseen, unsung,
   Until the morn grew old.

In gorgeous hues it came
     And cast its sheen
     Thy waves between,
Until the shadows crept away,
Abashed, beneath the touch of day,
   And died in shrouds of flame.

Oh, here, where dream-forms lie,
     And willows hide
     The gentle tide,
The soul could drift forevermore,
And learn of lake and rock-bound shore,
   Great thoughts that could not die.

FIRE-FLIES.

O'er the dusk-walls leaping
Where the shadows, sleeping,
    Darkly lie,
Comes a gleam of something golden,
Beautiful and bright and molten,
    Like the glimmer
    And the shimmer
Of the pointed star-lights,
    In the sky.

Upward from the meadow,
Close embraced by shadow,
    Soft wings rise;
Gold-sparks fill the haunted air,

STAR-DUST.

Sparkle here and sparkle there;
  O'er the river
  Gold-wings quiver;
E'er the night-wind, passing,
  Sobs and dies.

Are they spirits leaping,
Freed from death-like sleeping,
  By the night?
Are they souls we call immortal,
Free to pass life's mystic portal,
  In the gleaming
  And the dreaming
Of the magic-haunted gloaming,
  Starred with light?

Silent is the singing
Of the fire-flies winging
  Dusk-paths nigh.
All we see is something golden,

Beautiful and bright and molten,
  Like the shimmer
  And the glimmer
Of the pointed star-lights
  In the sky.

EASTER SONGS.

"With living lilies was the dark cross spanned."

THE FLOWER SONG.

Sweet peace that hideth in the lily cups,
    All brimming over!
Sweet joy that lieth in the pansy blooms
    And scented clover!
All earth, in lovely blossoming, her care
Doth lose in emerald sweeps and flowerets fair.

What means this splendor of the summer dawn,
    So pure and tender,
The spell that wraps the early violet pale,
    And crocus slender?
Ah, this—the gladsome earth hath found her king;
He lives—so whispers each sweet breath of spring.

He lives—oh hearts that bend to His dear will
    And break in sorrow,
For ye that mourn there dawns a brighter day,
    God's glad to-morrow!
In each bright Easter bloom for earth there lies
His glorious promise from the sapphire skies.

His peace! 'Tis written in the glad blue eyes
    Of violets waking.
He knows the voiceless pain—He heals the wounds
    Of earth hearts breaking.
He lives—upon His cross cast pain and tears;
His love is thine and mine through endless years.

THE BIRD SONG.

In the forests hear them singing,
    Sweet and clear,
Strains that angels, earthward bending,
    Lean to hear.

STAR-DUST.

Through the spaces, softly blending,
Earthly songs, to heaven ascending,
Bear the joy of life unending
    Through the sky.

THE EARTH SONG.

Christ is risen! Through the azure
    Floats the peace supreme
Unto souls for which He suffered—
    Suffered to redeem.
Christ is risen! O the glory
    Of this Easter-tide,
In its spring-time splendor sweeping
    O'er the earth land wide!
Earth is tuned to sweeter singing,
Bells and flowers to music swinging;
Through the skies her joy is winging
    Unto Christ, our king!

THE SONG OF THE BELLS.

I sat last night at my window
    As the solemn church-bells rang,
And I listened, with many a heart-throb,
    To the musical song they sang—
    To the beautiful sighing and swelling;
    To the tales of life they were telling—
    The story of Christ and his dwelling
        On earth for the glory of men.

And my heart was filled with the story
    Of Christ who for men had died;
As I listened, with faint heart swelling
    To the bells of the Easter-tide.
    And I wondered if souls, in their madness,
    Could list to this story of gladness,
    And cast away pain and all sadness
        In the peace of the Easter-tide.

O hearts that are weary of sinning,
    And hearts that are weary of life!
Would you list to the bells and their music,
    Your lives would be free from strife.
    Your souls with their peace would be thrilling,
    All the pain and the bitter grief stilling,
    With the rapture and joy swift-filling,
        Of the musical Easter bells!

O beautiful song of the Easter!
    There must be a God, I say—
A power supreme that is loving,
    And I cast all my doubts away.
    Let men all their creeds be forsaking,
    Their faith with philosophy breaking—
    I hold, to a heart that is aching,
        The peace of a God is supreme.

O beautiful bells of the Easter!
    Glad may your chiming be—

May you bring to all hearts that are breaking
    The peace you have brought to me.
May your musical sighing and swelling,
To the souls that are weary be telling,
The story of Christ and his dwelling,
    On earth for the glory of men.

THE VOICE OF THE LILIES.

The voice of the lilies I heard one day,
The voice of the lilies floating away,
    With a sigh and a sobbing,
      And a song that was sweet,
    Like a pulse that was throbbing
      In a circle complete.

The voice of the lilies I heard one day,
And this was the message they seemed to say:
    "In the dawn we are coming,
      In the night we will die,"
    And their song was a humming
      That died in a sigh.

The voice of the lilies I heard one day,
In the great silent churches it floated away.
   It rose with a sighing,
     Then sank at my feet;
   I knew they were dying
     These lilies so sweet.

The voice of the lilies I heard one day,
" For Christ we are dying," it seemed to say—
   " It is death to be giving
     Our incense and light;
   But far better than living
     Is to die for the right."

The voice of the lilies I heard one day,
In the great silent churches it floated away.
   'Twas the voice of all sorrow
     In each pain-freighted breast;
   'Twas the voice of God's morrow
     That bringeth men rest.

THE CHRIST-SONG.

I.

Christ is risen! Easter blossoms
   Bud and bloom anew
In the valleys of the spring-time,
   Arched by spaces blue.
In the flower-swept fields and meadows,
   Smiling, walks the Spring.
From her violet aisles the bird-songs
   Swell for Christ, the King.
      Mount to mount rejoices;
         Earth is passing fair!
      All her breath is incense,
      All her voice is prayer.

II.

Christ is risen! Easter song-bells
   Voice the love of God
As they swing with joy triumphant
   O'er the blossoming sod.

In the aisles of dim cathedrals
  Organ-notes of prayer
Domeward float on rapturous song-wings
  Through the incensed air.
    Slipping through the sunshine,
      Pinioned, dies dark care.
    Peace! the bells are ringing:
      Peace supreme and prayer.

III.

Christ is risen! Souls of mortals,
  Don the holy bloom
Of the spirit, heaven-ascended
  From the shadowed tomb.
On the pulsing heart descendeth
  Joy this Easter-day;
Lilied hosts, with golden censers,
  Christward point the way!
    Song to song is swelling;
      Men the joy-notes sing,
    Earth with rapture filling,
      Christ still lives, our King!

ECHOES.

In the sun a shadow fell—
      Was it spring?
Violet bloom and lily bell,
Gorgeous glow and radiant spell,
      Will it bring!

In the morn a wind swept by—
      Was it this
Waked the pansy, slumbering nigh,
From her blossoming dream-thought shy
      With a kiss?

In the noon a cloudlet passed,
      Sweeping low,
In the shadow, lowly cast,
Emerald buds crept thick and fast,
      All aglow.

In the eve a song crept by,
   Old yet new;
And the song-bird hidden high,
Sang of summer coming nigh.
   Was it true?

SNOW CROWNED.

Snow-crowned in April weather,
   The blooming vales below;
Hand in hand together,
   The flowers and the snow!
She cometh not with smiles and tears,
Sweet April-queen of by-gone years;
But lo! in trailing garments white,
As cold and pure as silver light.

No maid, with eyes down-drooping,
   And warm, soft cheeks aglow;
She wears no rainbow garments,
   But only robes of snow.
Her hands are cold; she brings no flowers,
To speed the swift-winged spring time hours.
Oh, April, fickle wert thou ever!
Shall we have faith in thee? Ah, never!

Poor queen, the frowning winter
   Hath chilled thy young sweet art;
Thou liest, pale and shivering,
   Upon the mountain's heart.
The southern breeze, thy lover bold,
Hath lost his warmth, grown wan and cold—
He hath no arts to make thee smile,
As was his wont, with gentlest guile.

The mountains are stern wooers;
   They give not smile or sigh
Their royal heads are lifted
   Against a cold, blue sky.
But April-queen, with sad, pale face,
Thou knowest yet some winsome grace:
Ere summer winds begin to blow
We'll find thy blossoms 'neath the snow.

JUNE.

JUNE.

Queen month of all the year,
   Crowned with the summer's splendor,
Thine eyes shed glory everywhere,
   Oh, June, so warm and tender!
So hath the full tide of the year
      Swept over thee —
      The summer's sea,
That sweeps away the frost and gloom
   From all the earth.

The days are full of joy,
   Marked by the birds' sweet singing,
Through forest glade and dusky grove
   Faint melodies are ringing.

The rose-blooms tinge the slopes with fire,
    Earth's loveliest flowers ;
    For June's swift hours
These fragrant blossoms bud and bloom
    In queenly state.

The star-lit nights how fair!
    The crescent moon appearing
Above the rim of mountains old,
    Their dusky cloud-ways clearing.
Her hand-maid, Venus, leads the way,
    Soft virgils keeping ,
    And, westward sweeping,
The star-flowers, like the roses, bloom
    And pass away.

Oh, June, born for the sake
    Of every dreaming rover,
Love hides in all thy roses red,
    And in the blossoming clover,

Sweet dreams float through thy woods and
> glens,
>> And life doth seem
>> A summer's dream,
Beneath the glory of thine eyes,
> Oh, queenly June!

AN AUGUST NOON.

O, emerald seas, full-tide with clover,
   And wind-swept grasses high,
How sweet, when August fans her embers,
   On your cool waves to lie!
No bark we need to stem this tide,
   To brave this golden weather;
Clouds, leaves, birds, bees, flower-sails and we
   Go drifting on together.

We hear the step of summer going—
   She walks with languid tread;
Trees bend to catch the opal gleaming
   That crowns her sun-touched head;
Through leafy spaces, flowered with blue,
   Far floats a world-wide splendor;
A song drifts down—the wild-bird's note,
   And dove-calls soft and tender.

O, silent trees, in silence speaking,
    And hills half lost in blue,
On pulseless noons, the heart, up-reaching,
    Finds FAITH undimmed in you!
Soft yellow buds and wind-swayed leaves,
    Fade, fade, ah, never, never!
O, clouds and birds and flower-sails,
    Drift on, drift on forever!

SEPTEMBER.

September, thou art dying,
   And dost thou know it, queen?
Oh, thou with crown resplendent,
   Oh, thou of lovely mien!
Upon thy hills the winter
   Lurks with his frosty breath;
September, dost thou hear him?
   He whispers of thy death.
Thy hands are toying idly
   With blossoms and with leaves,
Sweet harvest of the summer,
   Are these, thy golden sheaves.

First-born art thou of winter,
   September, child of light;
Thy face, of all, the fairest—
   Yet knowest thou the blight.

SEPTEMBER.

Thou rivalest the summer,
    With all her beauteous grace:
Yet in thine eyes, September,
    The shadow we must trace.
Thy cheeks flame with the crimson
    That hers hath never worn;
Yet, 'tis the fever flushing
    Of those whom death must mourn.

Thou mak'st me sad, September,
    With all thy life and glow;
As things we hold the dearest,
    I know thou, too, must go.
Thy splendid mounts will darken,
    Thy leaves will fade and fall,
The snow will come, September,
    And hide and keep them all.
Next year thou'lt come, September,
    But will mine eyes see thee?
More frail than thine, September,
    Our own short lives may be.

THE OLD YEAR.

Dying, dying is the year,
   And the earth is sad;
Sighing, sighing are the trees,
   And the winds are mad;
Creeping, ere the world be sleeping,
     Shadows drear
     Cross the year.

Dying, dying is the year—
   Old earth, do you care
For the child, now tired and sad,
   Once so glad and fair?
Dying, while the winds are sighing,
     Drifts of snow,
     Hide graves below.

THE OLD YEAR.

Dying, dying is the year—
   Fare thee well to-night;
Thou hast brought us smiles and tears,
   Shadows and the light;—
Fading, while the dusk is shading,
     Stars of light,
     From our sight.

Dying, dying is the year—
   Dreams we must forget.
Buried are the hopes it brought,
   Buried each regret;
Sleeping, waking, smiling, weeping—
     All the sad,
     All the glad.

Dying, dying is the year—
   Comes the new to-night,
Child of light, with wings of gold,
   Shadowless and bright;

Flinging clouds of joy, swift winging,
 O'er the past,
 Fading fast.

Dying, dying is the year—
 Let the sorrow die;
Bells ring out the sad, I pray;
 Winds, forget to sigh;
Sorrow, reign you not to-morrow,
 When the year
 New-born is here.

LOOKING BACK.

Why should we dream of vanished days
   And summer's glowing splendor?
Why should we mourn for swift-flown years
   And dream-loves warm and tender?

The shadow falls across the snow;
   The year is fading fast;
Our feet press on, the years wait not;
   Why should we mourn the past?

Each summer hath its crown of flowers,
   Each winter hath its snows,
And so life's circle grows complete
   With pleasure and with woes.

Time, in his endless, swift-winged flight,
   Brings to our hearts no rest,
Until we sleep the dreamless sleep
   In old Earth's pulsing breast.

The future oft holds happiness
　　And days of sweet surprise;
And it is best we turn not back,
　　With longing, wistful eyes.

And yet, how oft, oh, stern old Time,
　　When we have reached some hour
Of happiness, we wish that you
　　Might lose your fateful power!

And leave us to that dream of joy—
　　Yet soon we wake again.
Life hath no idle summer seas
　　Where hearts may drift from pain.

There is no joy untouched by grief;
　　Smiles vanish into tears;
Remorseless change is life's decree;
　　We may not hold the years.

And yet some precious gifts we keep,
　　Some flowers that do not die;
The year may go—we mourn him not
　　Beneath his wintry sky.

THE NEW YEAR'S DREAM.

I know not why this wintry night
    I dream of summer hours—
Of roses hid beneath the snow,
    Of sudden, swift-blown showers;
Of rains that softly sweep away
    The frowns from azure skies;
Of tender noons when, hushed in sleep,
    The dreamy landscape lies.

I know the flowers are dead and gone;
    They lie beneath the snow;
And where the sunshine used to pass
    The shadows come and go.
The purple hills are white and cold,
    The mountains, too, so lonely;
Yet, still, I do not think of these,
    But of the summer only.

In dreams I feel the roses' breath
    Upon my face—so sweet—
And scarce dare tread lest I should crush
    The violets at my feet.
I feel the hush of rose-red dawns
    And sunsets lost in splendor;
The summer world so filled with joy
    And song notes soft and tender.

And then I wake to find the gloom
    Of dark and dreary days,
I see the shadow on the snow—
    The sumbre wintry haze,
And wonder why the new year comes
    With storm and driving sleet.
The pure young year—its birth should be
    Among the May-days sweet.

And yet, dear hearts, I know full well,
    The flower-buds are but sleeping;

The summer gave, with loving trust,
    Them all to winter's keeping;
And they will wake to sweeter life
    Some day, when birds are calling;
So may our heart-blooms, hid away,
    Be not beyond recalling.

SUNSET.

Gold, gold, gold!
    Gold in the meadows of God,
    Gold in the blossoming sod,
Gold in the crown of the yellow sun.

Rest, rest, rest!
    Rest in the purple sky,
    Rest where the cloudlands lie,
Rest for the lands of the wind and sun.

Dreams, dreams, dreams!
    Dreams for the ones that we love,
    Dreams for the star-souls above—
Dreams for the world when the day is done.

IN THE RAIN.

As I rode last night through the city,
    In the dusk of a starless eve,
Such a weariness came o'er me,
    All my hopes could scarce relieve.
It came of the rain and silence,
    It came of the lonely night,
Like the shadow that follows grimly
    In the wake of the golden light.

Oh, the rain hath a spell of madness
    To bring back forgotten years,
And the falling drops have a murmur
    That accordeth well with our tears.
For the sins of the heart rise upward
    Like the mists of the dripping rain,
And the tears that are born of sorrow
    Well up from the heart again.

I fear not the flash of the lightning
    Nor the frown of an angry cloud,
For I hear the voice of the Infinite
    Speak in the tempest loud.
But, oh, how sad is the sobbing
    Of the wind in a dreary rain,
Like the wailings of restless spirits
    In the throes of an endless pain.

If ever the dead that are banished
    Come back to the earth again,
It must be their voices that mingle
    With the sob of the summer rain.
And if ever our hearts are longing
    For the sight of a vanished face,
Tis most in the rain and shadow
    And not in the sun's glad grace.

AFTER THE RAIN.

After the rain, the bow of peace;
    After the rain, the sun;
Gorgeous glow in the clouded west;
    Stars, when the day is done.

After the rain a shimmer of gold
    Shines in the crimson morn;
Blossoming blooms in the sun unfold,
    Emerald buds are born.

After the rain the world is fair,
    Sweet with the peace of God;
Fair from the mountain's snow-white crest,
    To the fertile, upturned sod.

After the rain, the shadow of pain
    Falls from our hearts away;
After the rain we lift our souls
    To the peace of a sun-crowned day.

THE AISLES OF PAIN.

The temple of God is fair and high,
Its altar builded of hope and sigh;
To heaven its corridors lead the way,
But ere we reach them we must pray
    In the aisles of pain.

To the stars uprise its spires of gold
From the mists of the ages, dark and old,
When the heads of kings in the dust bowed down,
And yielded scepter and yielded crown,
    In the aisles of pain.

And we who pass through the lonely night
From the depths of gloom to the walls of light,
Must kneel in the dust as lowly down,
And give up pleasure and honor's crown
    In the aisles of pain.

THE AISLES OF PAIN.

The aisles of pain are darkened with tears,
And stained with the blood of cruel years;
And the shiver and moan of crime and death
Go up to God with each throbbing breath
    From the aisles of pain.

The martyrs walked in the olden days,
With bleeding feet, through the narrow ways,
And we who follow must wait as they,
For the hand of Christ to lead the way
    Through the aisles of pain.

We may mock at pleasure and mock at pain,
And our lives may vanish in sun or rain;
Yet, soon or late, in the silent years,
We must kneel in sorrow and walk in tears
    Through the aisles of pain.

SHE AND I.

Just thirteen, and the book of life
   Scarce open lies at her feet;
Just thirteen, and her life is fair—
   No bitter in all the sweet.

Her hands reach out for the woman's joys
   With no thought of the woman's woes;
She longs for the sun that opens the bud—
   Nor thinks that it kills the rose.

She dreams that life is a pleasure field,
   And I—I wish it were true,
For the sake of this child with hands out-reached
   For the old, old truths that are new.

Our lives are near in the every day—
   And yet they are far apart;
She knows not the depths of our womanhood,
   And I—I have sounded my heart.

Dear child, though you walk with your hand in
    mine,
  There are terrible depths between;
You must kneel in the shadow and rise in the light,
  Ere you walk in my faith serene.

To grow means to suffer, one woman hath said;
  The lesson is true to us all,
And I would not turn back with you, dear child,
  Whatever your life befall.

THE CRESCENT MOON.

O'er yonder peak it sailed away,
    A bark upon the cosmic sea,
    The tideless ocean sweeping free
Through viewless spaces dark and gray.
    And all the bright worlds, soft and far,
    Each lovely, lonely, pointed star,
Swept westward in the silver trail
Of that fair crescent, clear and pale.

The old moon like a circle dim,
    Looked backward, faintly o'er the peak,
    Then followed, as a shadow meek,
The new moon's bright illumined rim.
    As oft our loved ones leave behind
    The dim sweet majesty of mind,
That, fading slowly, gives us grace
To bear the loss of form and face.

THE CRESCENT MOON.

How beautiful art thou, oh, moon!
    And yet a dead, dead world and cold,
    With silver peaks and rims of gold,
Thou knowest not one glorious noon.
    No dew-gemmed flowers, blowing sweet
    Grow at thy mountains' well worn feet;
No joyous birds with song-filled flight
Speed through thy realms of silver light.

Oh, crescent moon, that sailed away
    Through boundless seas unto the west,
    We long not for thy silver crest,
Nor for the dawns of Lunar day.
    Yet oft at night the burdened heart
    Yields to thy gentle, voiceless art,
And, seaking peace, finds impulse too
Towards all that's noble, good and true.

A SEA FLOWER.

A wave-tossed flower, clinging to the rocks,
    A gentle, tide-swept thing;
The opelet, with rose-hued wings out-spread,
    To dream, and sway, and swing,
In summer seas from dawn to starlit eve.
    Its petals are as frail and fair,
    As earth-land blooms that scent the air.
    Exquisite gleams of palest green
    It wears, and shimmering red between.
Ethereal blossom buried in the blue
    Of matchless seas!

Oh, rare and radiant flower—a monster thou,
    A foul and greedy thing!
Thy graceful arms do but invade the deep
    Its happiest lives to wring.

A deadly poison lurks within thy leaves,
    And they who dare draw to thee nigh,
    Frail creatures of the deep, must die.
    Satanic flower, as false and fair
    As sin art thou, whose garments rare
Are steeped in tears and blood and gilded o'er
    From hem to hem!

THE BLACK CAÑON.

The midday sun, in this deep gorge,
    Resigns his old time splendor,
His palace walls of dreamy gold,
    The rose-hues warm and tender.
        The cleft is dark below,
Where foaming flows the Sumbre river;
The wild winds sigh and blossoms shiver,
And violet mists ascending,
        Obscure the Orient glow.

O! rushing river, emerald-hued,
    How mad thou art and fearless,
No frowning gates, though granite barred,
    Can curb thy waters peerless!
        The silent gods of stone

Revoke their ancient laws of might
When through the gorge, with wing-swift flight,
Thy wind-tossed waves are speeding,
    Each moment wilder grown.

The faint stars shine in broad midday
   Through twilight mists, gold-rifted,
Where opal streams make dizzy leaps
   O'er jasper walls blue-rifted.
     Below, no naiad's dream.
'Neath dim arcades, through sunless deeps,
The nomad river lonely leaps,
Where castled crags rise skyward
     Like watch-towers o'er the stream.

On massive cliff-walls Nature's hand
   Has turned time's sun-worn pages;
In faces carved and figures hewn
   We trace the work of ages.
     The gold-tipped spires sublime

That pierce the sky like shafts of light,
But mark the measureless, heavenward height
Of Nature's own cathedral,
  Whose stern High Priest is Time.

In this grand temple, æons old,
 Her organ notes are pealing;
In gold-flecked arch and wave-worn aisles
 The flower-nuns are kneeling;
  Her altars echo prayer,
And when at dusk the cold moon shines,
O! awful are the far white shrines,
From earth to God upreaching,
  Through spirit-flooded air.

TO A DEAD BIRD.

Thou art no larger than my hand,
   Yet life once throbbed in you,
A life that loved each trembling leaf
   Beneath the arch of blue.
At daybreak—you were strong and warm,
Without a thought of shadowing harm;
At noon—you quit your emerald swing,
And lie at rest—a poor dead thing.

What is the grace that went from you—
   This Something which is gone?
A breath that moves the world of men,
   Yet transient as the dawn.
Oh! can we boast of God-like skill
And sciences, when still—oh, still,
The keenest eye must pause at this—
The veil of death—the soul we miss?

SUFFER AND BE STRONG.

I love to think this grand old earth
   Hath suffered æons long of woes,
Though now she wears upon her face
   The smiling beauty of the rose.

I love to think her trembling heart
   Beat fast in centuries of pain,
Ere God swept with His holy peace
   Her every lovely mount and plain.

In circling fires her heaving breast
   Was fiercely wrapped, for time untold,
And tempest-tossed she fought for life,
   While seething oceans round her rolled.

In darkness, like a thing accursed,
   Her form was whirled through spaces vast;
Yet, far beyond, a Hand was there
   That reached its might and held her fast.

And now she swings through starry realms,
  Obedient to a higher Will;
In glorious light her days are passed,
  Her throbbing pulses now are still.

Oh! by these mountains upward hurled,
  And all the earth-walls rent and torn,
We feel the woes that she endured,
  And all the anguish she hath worn

Yet on her breast the flowers lie,
  The peaceful blossoms of her love;
And oceans blue as summer skies
  Look up to stars that smile above.

Oh, heart, learn of thy mother earth
  To suffer bravely and endure,
If thou wouldst wear the crown she wears,
  If thou wouldst noble be and pure.

And though the pain be worse than death,
   And woes seem darker than the night,
The flowers will bloom along the path,
   Thy soul will seek and find the light!

OUR KING.

There lived a King in the olden days
   In the beautiful, warm-hued East;
But he wore no crown, and none sat down
   In his palace halls to feast.

He was fair and strong and gentle of heart,
   And he ruled by the right Divine;
But none came nigh, in the days gone by,
   To drink of his golden wine.

He wore no purple, this King of old;
   His only jewel was Love,
Yet fair and far, like some gold-winged star,
   It shone on his breast above.

His gold was Faith, and his throne was Truth,
   He was born to rule the earth;
Yet men came near, with jest and sneer,
   Denying his royal birth.

He was born at night in a lowly place,
    Where never the rich came near;
Yet the golden star, o'er the manger bar,
    Resplendent shone and clear.

And the wise men followed that guiding star,
    And the shepherds kneeled at his feet;
And seraphic song from the heavenly throng
    Awoke for him music sweet.

There lived a King in the olden day,
    And no prince descends in line;
Yet he rules to-day, and he rules alway,
    By the right of his love Divine.

Oh, ring, sweet bells, on the Christmas dawn,
    And this be your tidings then:
Joy, and mirth, and peace on earth;
    Peace and good will to men!

SONGS OF LOVE.

L'AMOUR.

By the silver heights he sat and sang,
  This beautiful god of old;
He, with his fair hair backward tossed,
  He, with his harp of gold.
    And the sun by day
    And the stars by night
    Heard the trembling strains
    From the cliff's far height,
Like the words of a song twice told.

His name was Love, and he wooed the World
  From his silver cliffs on high,
But she turned away from his golden harp,
  And scorned his beautiful lie;

For she knew full well
    That her maidens fair
From the cup of Love
    Would drink their share,
Ere the song god passed them by.

And her dream was true, for the maids of earth
    At the sound of his song drew nigh,
And they knelt at the feet of the sweet-voiced king
    And worshiped his golden lie,
        Till Time crept past,
            With his dark-hued years,
        And prisoned them fast,
            With a chain of tears,
To the silver cliffs on high.

On his flower-decked throne he still doth sing,
    This far-famed god of old;
But he gives no thought to the women fair
    Who follow his harp of gold.

L'AMOUR.

  And the World will pass
   And the new grow old,
  Ere the sad, sweet words
   Of his song are told,
And the breath of his lips grows cold.

STAR-DUST.

BRIDAL VEIL FALLS—YOSEMITE.

The stern old monarch hath claimed his bride,
The beautiful maiden who stands aside,
And hides her face in the veil of snow,
That falls like a mist on the rocks below;
But her love is not for El Capitan, old —
He, with his grand face carved and bold;
And she turns her face to the wall of gold,
Like a weeping child with a grief untold.

The shimmering veil in the soft breeze sways
To and fro in the golden haze,
Turned by the tints of the amber sun
To a thousand hues ere the day is done.
But the rainbow colors reveal no trace
Of the Indian maiden's living grace;
And she stands in stone carved evermore —
A type of love in the legend's lore.

BRIDAL VEIL FALLS.

Patient and grim the bridegroom stands,
Forever waiting with empty hands
For the beautiful bride who turns away,
Hiding her face in the white-foamed spray;
And the evil wind with a spirit of hate
Points to the high-domed walls of fate,
And mockingly wraps in his own embrace
The white-robed bride in her veil of lace.

Oh, men will love and women will hate,
And high and grim are the walls of fate,
But the spirit of God in its forms untold
Will be written forever on walls of gold,
And they who worship at Nature's throne
Will read His truth in the carved white stone,
The grand, true words on the priestess' door —
The grace of love in the legend's lore.

TO A ROSE.

I look in your heart's deep shadows,
    Where the fiery crimson glows,
And your cheek where the sun has kissed you—
    My beautiful, queenly rose;
And I long for your fire and beauty,
    And your glorious, sunlit ease,
I long for your spell and glory
    And your power to give and please.

You, who are queen of the flowers,
    Have all the world at your feet,
And I envy your crown and sceptre,
    And long for your kingdom sweet.
You have the sun and the garden
    And the lake where the fountain plays;
I have the heart of a woman —
    And a woman's patient days.

TO A ROSE.

Deep in your heart is the passion
    Of love and its power divine,
For the sun has loved and caressed you,
    And his glory and love are thine;
Yours is the mission of beauty,
    And the power to give delight,
And I envy your perfumed graces
    And your spell in the dewy night.

You are the loved of all lovers,
    The symbol of truth and grace,
And I long for your crimson glory,
    And your beautiful, sun-kissed face.
For the night and the stars are yours,
    And the throne where the sunlight glows —
While I have no king or kingdom,
    My beautiful, queenly rose.

But know you this, you proud one!
    Yours is a swift, dark death;

I could crush the life from your petals
    With one quick, sobbing breath.
Mine is the life that liveth,
    Sprung from the grace of God,
For I am a soul immortal,
    While you are born of the sod.

THE OLD AND THE NEW.

From the mists of the past, a legend I glean
   That is old as old can be,
And a fragment of song that is sad and sweet
   That comes to me over the sea,
Of a life that was lived in the olden day —
But life is the same, forever I say.

In the olden days there were kingly kings
   And knights who were brave and true;
And women who loved, not just for a day,
   But oh, for a life-time through.
And men who braved death without thought of fear
For a woman's smile or a woman's tear.

The dusty volume is laid aside —
   Its pages seem ages old,

But somehow my eyes with sorrow are filled
   For the sad old story it told.
For the song and the sorrow have passed away,
But life is the same, forever I say.

In the olden days this maiden lived,
   This maid of old — who was fair,
And the shadows of night were in her eyes
   And the dusk was in her hair.
Behind her the old year sank to sleep
In the ocean of Time, infinite, deep.

And a spirit born of the night's dark grace
   Swept through the shadowy air,
Prisoned the gold in his dusky wings —
   Prisoned the stars so fair,
He came from the moon and he came from the sun
To shadow the glow of the year that was done.

In his arms he sheltered the new-born year —
   And oh, 'twas a precious thing!

But what was this child with its life unlived
   To the year that was dying, a king!
And the maid looked not to the new year fair
But she turned to the old, for her heart was there.

"Choose!"—and the spirit touched her hair—
   "Treasures this year doth hold,
Riches and joy!" But the maid replied,
   "Spirit! I choose the old;
For my heart is there and 'twere sweet to lie
At rest with the year—if he must die!

"Old things are best and I love the old,
   Though it brought me less pleasure than pain.
Old things are best, and I love the year
   With its measure of sun and rain;
For it brought me love, and oh! I could die,
For that measure of love—were it even a lie.

"I crave not the treasures that thou would'st give,
   Life has but one love, I say,

And the pleasure and pain that he brought to me
   Will be mine, forever and aye.
Oh spirit of light, were your year of gold,
For the love that it brought, I would choose the old!

" For he came to me in the sad old year,
   And left on my lips his breath;
And I gave him all that my heart could give,
   And better, without him, were death.
For no other kiss on my lips shall lie,
Than the one he gave — for that I could die.

" But, Spirit of Light, one boon I crave
   Out of your priceless store:
The new year take to the lover I loved,
   And give him the burden I bore.
Oh, lay on his heart the burden love brought,
And the pain and the tears his kisses wrought."

In the olden days they could die for love —
   And she turned from all men apart,

With the shadow of love in her dusky eyes,
    And the shadow of death in her heart.
Asleep in the ocean the year did lie.
And is it for love that a maid must die?

The spirit smiled as he winged his flight
    Over the sea and land,
And he touched with pain the lover's heart
    With the new year's magic wand.
Oh, she died for love in the olden day —
But love is the same, forever I say.

And so the old legend is sad and sweet,
    With a sadness that is but true,
For the years will come and the years will go,
    And maidens their love will rue;
And the song and the sorrow will pass away—
For life is the same, forever I say.

A DREAM OF THE EXPOSITION.

Gilmore played — it was autumn weather,
   And the air had a tinge of frost;
But she looked like a beautiful rose of summer,
   In the gold-swept forests lost.
Her eyes were dark, like the proud Sultana's,
   Her lips, like the leaves, were red,
But the glory of glories, the crown of her beauty,
   Was her wonderful, sun-touched head.

HE.

She sat near the aisle where the sunbeams slanted;
   'Twas foolish I know — and yet,
It seemed to me all the gold rays flashing
   On that beautiful figure met.

Gilmore played — ah, yes, it was music
   Fit for the gods of old;

But my ears scarce heard, and my eyes seemed blinded,
   By the quiver and flash of gold.

We had waltzed in the gorgeous hall of the Prophets,
   Near the fountain of glittering spray;
She had looked in my eyes—was I wise to think it?—
   In a womanly, tender way.

But why should I claim? perhaps some other
   Was won by that same sweet glance;
I only knew that my pulse still fluttered
   With the thoughts of the dreamy dance.

She had looked like a queen in that velvet court robe,
   With the billows of filmy lace,
And that wonderful, high standing Queen Anne collar
   That framed in her fair young face.

Was it "William Tell" the band were playing?
   She said so; she ought to know.
I remember only the smile she gave me,
   And her eyes, with their shaded glow.

Too soon was the dream of the music ended;
    Too soon were the pictures read;
I know not why, but my life seemed ended
    When the white lights were lit overhead.

For she turned away, with her grace and beauty,
    Her "Good-bye" was proud and cold.
Ah, this is the way women have of refusing
    Our love ere it even is told.

Ah! not for me is this sweet, sweet vision
    And foolish to dream was I.
Is it well, I wonder, to care for a woman,
    Or long for her smile and sigh?

SHE.

I thought at the ball he looked so handsome;
    And the waltzes—they seemed divine;
And he said — but, oh, how these men do flatter! —
    The fairest of faces was mine.

He looked in my eyes; oh, surely he found there
  The loving I could not hide;
But he spoke no word, and I hid the longing
  'Neath the mask of my woman's pride.

We met again when the band were playing;
  I smiled when he touched my hand;
But I turned away when the music ended,
  For he seemed not to understand.

'Tis a woman's lot to love and not tell it;
  It is bitter, I know, and yet—
I would rather die with that love unuttered
  Than grieve o'er one spoken regret.

L'ENVOY.

Gilmore played—it was autumn weather,
  And the air had a tinge of frost,
And these two apart, yet in heart together,
  Mourned what they both had lost.

Was it fate that had come between them, decreeing
    That the words should never be said?
Was it better thus than to stand together
    One day, by the grave of love dead?

THE WALTZ-QUADRILLE.

Oh, the dreamy spell of the Waltz-quadrille—
   To sway, like the wind-tossed flowers,
To the rhythmic beat of the music sweet,
   In the joy of the flying hours!

Forward and back the light feet go,
   Moved with the spirit's gladness;
No thought of care in the bright hearts there,
   Forgotten the world of sadness.

Yet, why should there come, in an hour like this,
   The spell of a past, dead sweetness,
The grace of a day that died away
   In the summer's joyous fleetness?

The pulses stir with a dreamy pain,
   A mist dims the warm light's splendor,
As the dancers pace, with a swinging grace,
   To the waltz-strains, low and tender.

There's a scent of roses down the hall—
    A languor that half entrances —
And a dream of love in the eyes above
    That droop 'neath the answering glances.

But hands that are touching and hearts that throb,
    With the rapturous music swelling —
Do you think it is new, with no pain or rue,
    This joy you are silently telling?

There's a spell in the hour that's half divine,
    But the morning brings cold forgetting;
The smile and the sigh with the music die,
    And the sorrow of life is regretting.

Oh, the memories bright of the Waltz-quadrille,
    And the dream of the joy-filled hours!
They touch the heart with a tender art,
    Like the grace of the dead, sweet flowers.

ORANGE BLOSSOMS.

Do you dream of a wedding, sweet maiden fair,
You, with the blooms in your dusky hair?
Are your thoughts of a lover so proud and tall,
That you blush like the rose on the garden wall?
      'Tis a dream of the night,
      And the blossoms white
    Are the stars of a heaven to you.

Do you dream that the pure, white flowers of love
Will shine for aye as the stars above?
Is the crimson flame of your love so bright
It can shine through the shadows of love's dark
   night?
      If this be your thought
      Then your life is wrought
    With the spell of a passion true.

Are your dreams of a lover so fond and true
He will keep life's sorrows forever from you?
And the sun of your love, will it shine alway
As it shines on the morn of your wedding day?
      In your dusky eyes
      The glad truth flies,
Like the flash of a holy light.

But the blossoms will fade in the brooding eve,
And the shadows will fall and the stars will leave;
Life is no dream, and the flowers so fair
May herald the coming of pain and care,
      And love's pure prayer,
      Like the blossoms there,
May die in a starless night.

"THE MERRY WAR."

We sat at the Cave — Do you remember?
   Mid-summer, one year ago;
The gaslight flickered o'er pallid grasses,
   As the moonlight over snow.

The past comes back like a dream forgotten;
   They were playing the "Merry War";
You caught the flash from a woman's weapon —
   I think there is still a scar.

You thought of that blonde and the high soprano;
   I thought of the flower, dead —
The flower you gave me, tender and fragrant.
   "The tenor was fine," you said.

The music rose and swelled to a rapture;
   The singers were at their best;
But I thought of the flower, faint and withered,
   That lay upon my breast.

You had touched it once, with careless fingers;
   "A passion flower," you said,
With a tender smile and a look that thrilled me—
   The blossom was crimson-red.

And yet you looked at that blonde girl near me,
   With your soul within your eyes.
She turned to me — her look was a challenge:
   It was war for a paradise.

Her face was fair with a wond'rous fairness.
   Do you wonder my heart beat fast,
That the red blood leaped to a sudden madness,
   For a joy that was almost past?

Her pale, gold hair, in the gaslight shimmered,
   Her smile cut my heart like a knife;
I laughed, but I knew that the warm blood, ebbing,
   Was taking away my life.

"The Merry War"—oh, the music, thrilling,
   In my dreaming haunts me yet;
And the pale young moon that was slowly rising
   After the long day's fret.

You encored the tenor—yes, he sang sweetly;
   But the flower lay dead on my breast.
You looked at her, with that new love thrilling—
   Do you care to hear the rest?

She won—why not? She had riches and beauty;
   I had only my love for you.
She held the wine to your lips. You drank it;
   But the draught, it was bitterest rue.

Men play with love in a curious fashion,
   But true women scorn a lie;
And the one who won you with such deception
   I'd rather were she than I.

We sat at the Cave — Do you remember? —
   Mid-summer, one year ago.
Were we there to-night, would she win, I wonder —
   That blonde in the gaslight's glow?

THE PAINTER'S MODEL.

I see her before me in the light—
    A beautiful work of art—
With the sun on her cheek and her red gold hair,
    And the cross above her heart.
Is it a nun I paint to-day,
With her sombre dress and eyes of grey,
    Or a woman reaching higher,
    With her throbbing pulse of fire,
And her heart like mine,
Filled full with its measure of life's red wine?

Her hand slips out of its sombre sleeve—
    There's a glimpse of an arm divine,
A fair white throat, and a face above,
    That is perfect in every line
Was there ever a nun I wonder like this,

With a red, sweet mouth that was made to kiss?
    By all that's sacred, I swear,
    That fair nun sitting there,
Is fitter to love,
Than to wear the plumes of a convent dove.

Ah, well—what folly is this of mine,
    To work now, man, with a will,
And catch that exquisite, saintly smile,
    As she sits there so calm and still,
With her beautiful red lips just apart,
And the sun on the cross above her heart
    Know this, oh, my fair-faced nun,
    With your head in the golden sun,
To me you'll bring
The echo of fame—'tis a glorious thing.

But lo! I swear that my eyes are dim,
    And my hands do tremble so
That I lose the power to catch the shade,
    And the exquisite, golden glow;

And, by all that I love, I value my art
Far less than the cross on her throbbing heart
  In this moment my arms drop down,
  And I hold, oh Nun, that the crown
    Of fame is a thing
Senseless to the pleasure that love must bring.

She moves—and the spell is broken quite.
  A woman I see—not nun.
Her cheek is tinged with the wine of love
  That leaps in the golden sun.
She is the work of God's highest art—
A woman—with the cross above her heart.
  And shall I refuse the boon
  He gives? Like the stars and moon,
As free and fair—
That exquisite saint with her red-gold hair!

PALMISTRY

They sat in the glow of the crimson light:
    She, with her face bent low,
While he, with his proud, imperious smile,
    Imprisoned her hand of snow.
And he read the lines with a grave delight:
    For the signs were so fair to see;
And the slender hand was as curved and fair
    And perfect as any might be.

It has grown to be a fashionable thing,
    This study of palmistry;
And men and maidens had studied the art,
    Just to see what their fates might be:
And lo! in a trice, so wondrous wise
    Had these students of learning grown
That unto their science was nature revealed
    And the secrets of art made known.

And now in the twilight he sat at her feet
   And held in his hand her own,
And thought that the life-lines written there
   Were the fairest he ever had known;
And she, who knew naught of his magic art,
   Looked down on his lifted face,
And dreamed that the smile in his proud, dark eyes
   Was the symbol of love's sweet grace.

Oh! the beautiful rose, with its crimson heart,
   Was never more fair than she;
And he knew that the slender hand he held
   Was perfect as any might be.
What wonder, then, that he read the lines
   By the light of a new-found art,
And felt that the science of palmistry
   Had spoken the love of his heart?

In the shadowy gloom of the purple eve
   She sat with her face bent low;

And he, with his proud imperious smile,
  Imprisoned her hand of snow.
But not for the sake of palmistry
  Did he look on her glowing face:
But all for the sake of the love he bore
  And the charm of her tender grace.

Ah! the gypsy lore, in the days gone by
  Was wondrous and very wise;
But more wonderful still is the art that reads
  The love in a maiden's eyes
The lines in the beautiful hand he held,
  In the twilight, he could not see;
But lo! by the touch of his master hand,
  She knew what her fate might be.

STRANGERS YET.

    "Shall we never fairly stand,
    Soul to soul or hand to hand,
    Are the bounds eternal set,
    To retain us strangers yet?'
You sang the song — I listened, silent;
    The words I can't forget;
Our hands had touched, with pulses thrilling,
    You whispered, "Strangers yet?"
One bright star, smiling, hung above us;
    I hear the music still,
It brings me back the pain and longing,
    The sweetness and the thrill.

The dreamy waltz we danced together,
    We talked of fame and art,
And yet it seemed to me you counted
    The life-strokes of my heart.

STAR-DUST.

With languid step we paced the German,
    And smiled when it was through,
Then stepped aside from all the others —
    Our world was only two.

The Renaissance we talked of lightly,
    Our words were rambling, few;
Each might have talked in Hebrew, Latin,
    For aught the other knew.
Ah, we were measuring life by heart-throbs,
    We played a dangerous game,
That lost or won, the past or future,
    Can never be the same.

I listened to your tale of travels,
    For you had wandered far;
You'd looked for me the wide world over —
    We met beneath that star.

STRANGERS YET.

Ah, strange — my feet had wandered blindly,
    And found the way to you;
Once met — in heart's swift recognition,
    Each one the other knew.

Beneath that star you sang; I listened;
    There came a vague regret,
A pain and sweetness strangely mingled,
    For we were strangers yet.
Yes, strangers as the world would have it,
    We knew 'twas true, and yet —
We must have known each other always,
    Although we had not met.

How else can stranger hearts so quickly
    Each other understand,
Two souls that need but this dear language,
    The touching of a hand

There was no need for words that later
    You, trembling, spoke to me.
The heart that gives, the heart returning,
    Can never more be free.

The one star hung above the river,
    The one song died away,
Fate lifted up her walls between us,
    And this is just, men say.
The song you sang was too prophetic,
    And so that vague regret —
The months have passed by weary laden,
    And we are strangers yet.

'Tis wrong to wonder why and question,
    It must be best we say,
Yet oh, 'twas hard to find each other —
    Then go the other way.

The one lone star — the music thrilling,
   We never can forget.
Yet we must live, forever parted,
   For we are "strangers yet."

TWO YEARS.

1.

THE OLD YEAR — MIDSUMMER.

A rose is in her braided hair,
    A rose is on her breast;
Oh, bloom of blooms, as fair as yours,
    The ways her feet have pressed!

Summer ways and fields of clover —
    Love is like the rose;
In the grasses, full seas over,
    Love the flower grows.
Lovers part, yet glad the morrow,
    Where the soft winds blow;
Love is sweet — yet full of sorrow,
    Those who love must know,

TWO YEARS:

The year has budded, bloomed and faded,
    And sere the rose-leaves lie,
In dusty leaf-drifts, autumn shaded,
    The flower love must die.

II.

THE NEW YEAR — MIDWINTER.

A diamond in her braided hair,
    And diamonds on her breast;
Oh, jeweled wealth, your gifts are rare,
    Yet are your treasures best?

Marriage bells and bridal flowers,
    Wealth is like the star,
Golden-eyed, with matchless power,
    Life to make or mar.
Brides are fair — yet dark the waking —
    Memories that are past
Doom the heart to ceaseless aching —
    Hearts that love has claspt.

The year has budded, bloomed and faded —
    Oh, year of years just born,
In your bright day-dreams, jewel-shaded,
    Will love forever mourn!

PARTED.

You have kissed my lips, and yet we are parted—
   And the world, love, lies between;
Those lips are dumb, yet my heart still quivers
   With the pain so swift and keen.

My hands you have clasped, with love's dear thrilling,
   In the circle of joy complete;
And oh, for one of your dear caresses
   I could die, love, at your feet.

You have put me out of your life forever,
   And you called me, love, your queen;
You have doomed our lives to a bitter silence,
   And the world, love, lies between.

For us there will never be a meeting,
   We walk in the shadows apart;
Did we dream we could live and yet be parted
   When we stood, love, heart to heart?

But I know, one night, you will hear me calling
   In your dreams, love, far away,
And will start then, love, and hearken and listen
   To the words that I will say.

You will feel my breath on your dear lips falling,
   You will open your eyes and start,
You will clasp my hand with the old love thrilling,
   You will hold me, love, to your heart.

On my dusky hair you will press your kisses,
   You will look in my eyes and smile;
We will live again in the old sweet heaven,
   Oh, love, for a little while!

Our lips will forget their bitter silence
   And the dead days, love, between;
You will kiss from my eyes their tears of sorrow
   And the pain, love, swift and keen.

You have doomed our lives to the pain unending,
   With my words, love, all unsaid;
But I know, one night, you will hear me calling
   Like a voice, love, from the dead.

THE GOLDEN GATE.

We stood by the Golden Gate,
  He and I,
And the pale gold stars were late
  To come in the twilight sky.

He held my hand in his own;
  "Love," he said,
"Will you sail with me alone
  Over the seas of red?

"Is love as great as the sea
  Dreaming there?
Can you leave the world for me,
  Love, with the dusky hair?"

My head was against his breast;
  I made reply:
"I have found on your great heart, rest,
  For you will I live or die.

"With you will I sail the seas,
  Far or near;
In the dusk is my heart at ease
  With your dear presence near."

We sailed through the gate at night,
  He and I,
And he steered our boat by the light
  Of the late stars in the sky.

We sailed to the Isle of Love;
  "Dear," said I,
"My heart will rest like the dove
  In the land of love, till I die."

MY IDEAL.

He must be noble, and kind, and kingly,
    With a heart that is full of trust;
He must be gentle, and tender, and loving,
    With a will that is firm and just.

His eyes may be brown, or blue, or hazel;
    His form may be tall or short;
But the grace of love must he wear unbidden,
    Ere he worship at my love's court.

He must be brave, and true, and knightly,
    With patience to love and wait;
And he must be strong enough to carry
    The burden of love's dear weight.

He need not wear the crown of a poet,
    Nor philosopher's garb of thought;
But he must be swift to catch emotion,
    With a mind that is purely wrought.

He must have reverence for every woman,
    Because he hath love for me;
He must be gentle to little children,
    And kind to the old we see.

He must have courage to bear our sorrows,
    As we walk through the aisles of pain;
He must be glad of the sun and flowers,
    With smiles sometimes for the rain.

He need not tarry for conquering riches,
    For hearts are not won with gold;
But oh, with love must his ships be laden—
    With the riches of love untold.

If thus he comes, I shall know and love him,
    And all that is mine to give
I shall fling at his feet, my king, and serve him
    As long as we both shall live.

THE SUMMER'S NOON.

A full-tide sweetness fills the air,
   A fullness born of ripened grain
And bursting blooms; completeness rare
   That leaves an inner sense of pain,
Because the year's high noon is reached
   Whence all things change again.
The globe of dew, the star-like flower
   Reflect the universal laws.
Earth hath no bright, enchanted hour
   When life, complete, unchanged, can pause.
Then why so weak, oh, joy-swept heart, to dream
   That love can reach its fullest tide,
And know no waning autumn, fire-tinged,
   When summer's passion flowers have died?

GOOD-BYE.

The velvety bloom on the rose is gone;
   The sweetness of love is over,
The shadow falls on the crimson day
   And the dusky, dewy clover—
      Good-bye, dear heart, good-bye.

The summer of love is gone so soon—
   The summer of love together;
For lovers must part in the autumn-tide,
   Love dies in the wintry weather—
      Good-bye, dear heart, good-bye.

Hearts break when the blush of the rose is gone;
   Hearts break when the summer is over;
And only the wind and the falling leaves
   May echo the song of the lover—
      Good-bye, dear heart, good-bye.

GOOD-BYE.

Oh, better the long, sweet slumber of death
   Than the breaking of hearts asunder;
Oh, better than life with an aching heart
   Is the sleep of the dumb dead yonder—
     Good-bye, dear heart, good-bye.

Perhaps I, too, shall rest ere long
   In the slumber of death, low-lying,
But know you this, dear one, my heart
   Would love you even in dying—
     Good-bye, dear heart, good-bye.

LOVE'S RETROSPECT.

The violets slept on my breast, sweetheart,
   And your soft eyes held their hue;
Violet youth with its budding spring
   Was touched and revealed by you!
It was not summer, nor was there snow,
For nothing of seasons do lovers know.
Dear, I was all women and you all men,
And we lived all life in that timeless Then!

ON THE WILLAMETTE.

The beautiful river, so clear and wide,
   Was starred with a thousand lights
From the city's lamps and the soft star gems
   That symbol the sweet June nights.
      It was grandly fair,
      This river so rare,
   With the song of its music glad.

I dipped my hand in the sobbing waves
   Of the river so dark and wide,
And I caught the shadow of awful things
   That the rivers of mystery hide :—
      The hushed quick cry
      Of the souls that die
   In the rush of the river mad.

But the silver spray of the star-swept waves
   Banished such thoughts away,
And I heard no more the sad, low words
   That the winds and the waters say —
      The sob and shiver,
      The hush and quiver
   Of the reeds on the marshy shore

Oh! that the spell could last forever
   Of the music that throbbed and died —
The dream of the echo-haunted river,
   The song of the star-lit tide!
      But the stars and the song
      To the night belong —
   She holds them forevermore.

MINE.

They are mine — the seas and the mountains,
    The plains and the pine-clad hills ;
They are mine — the forests and flowers,
    The rivers and rippling rills.

You may take the gold and the silver,
    The riches and honors of earth ;
I'll claim Nature's simplest treasures,
    And the earth that gave them birth.

They are mine — the stars and the bird-songs,
    The waves and the glittering sky ;
They are mine — the shadowy streamlets,
    And the deeps, where the sunbeams lie.

You may have the crown and the scepter,
    The triumphs of power's sway ,
I'll take the night and her shadows,
    And the glorious, sun-lit day.

They are mine — the thoughts of the masters,
    The works of the good and the true;
They are mine — the songs and the poems
    Of God's kingly, chosen few.

You may claim for your own, life's pleasures,
    The feast and the crimson wine;
I'll take for my share art's treasures,
    With the truth in each golden line.

But the glory of God's sweet giving
    Is not that He giveth me,
But He giveth to all His treasures,
    And enough there will always be.

FORGETTING.

What is this art men call forgetting?
    Is it learned in a day or year?
Does the heart e'er forget a pain or sorrow,
    Or a memory once held dear?
Do the eyes that have wept forget their weeping,
    And the lips that have spoken grow dumb?
Do we leave the past and remember only
    The days that are yet to come?

Do we stand in the sun and forget the shadow,
    Or remember only the rain?
Do we clasp our joys in a tender passion,
    Forgetting the swift, keen pain?
Do our pulses throb with their own great rapture,
    Mad with the joy of life—
Forgetting the pangs of the countless millions,
    Doomed to unending strife?

Do we love till our hearts with love are aching,
   And forget that love in a week?
Do we grasp ambition with idle fingers,
   Forgetting the fame we seek?
"Forget!" cries the man in his careless fashion,
   "Forget!" says the woman of pride;
But the man still clings to the old sweet passion,
   And the woman weeps aside.

I tell you, hearts, there is no forgetting!
   You may laugh remorse to scorn;
You may hide in a grave your sin or sorrow—
   Each day it is newly born.
"Forget!" says the world in its heartless fashion,
   "Forget," as the men pass by;
But no man forgets as he carries his burden—
   We cannot forget till we die!

PANSIES.

"Pansies are for thoughts," they say,
    Beautiful and true,
Purple-breasted, golden-hearted,
    Flowers of royal hue:
Thoughts that come like sunset dreams
    In the dewy gloaming,
Dove-like, with soft fluttering wings,
    Through the wide world roaming.

Into hearts so deep and wide,
    Wondering I look,
Shell-like, do these flowers echo
    Songs of sea and brook;
Do they hold our inmost thoughts?
    All their petals filling,
Like the heart-beats, never heard,
    Through the silence thrilling.

They have caught the sunset tints,
    Pansies, bronze and gold,
And the star-light faint embroidered,
    In each purple fold.
Hold they, too, our dreamlike words—
    Good-byes, softly spoken;
Are they messengers of hope,
    Each fair leaf, a token?

Passionless these pansies are—
    Voiceless as the earth;
Yet the soul their beauty greeting,
    Wakes to nobler birth.
"Pansies are for thoughts," they say,
    Beautiful and true;
May their blossoms bring sweet thoughts,
    Friends, to each of you!

GOLDEN-ROD.

Beautiful golden-rod!
Up from the half-burned sod,
When the August fires have died away
You rise, gold-tipped like the sun's last ray,
Born of the summer's after-glow,
Reaching heaven, though prisoned below:
Beautiful golden-rod!

Fire-kissed golden-rod!
There, where the grasses nod,
Waving their yellow plumes of death
Touched by the frost-king's keen, swift breath,
Your slender spires of reddened gold
Like the meadows grown brown and cold:
Fire-kissed golden-rod.

"Lo! pass under the rod!"
Soft-spoken words of God.
One by one through the long-life day
Strewing with tears the grief-grown way—
God's erring children pass slowly by,
Stilling the heart with their passionate cry:
Lo! pass under the rod!

Beautiful golden-rod.
Chosen flower of God!
Under the shade of your spears of pain
Many a heart in the dusk has lain;
But the cruel barbs wear the dust of peace
God's love — it bringeth the sore heart ease:
Beautiful golden-rod!

AUF WIEDERSEHEN.

Auf wiedersehen, friend, I speak it,
   And tears dim not my eyes;
The German word hath sweeter meaning
   Than all our cold good-byes.

It hath a faith for future moments,
   And days that pass us by,
And so I say, *auf wiedersehen* —
   I will not say good-bye.

To meet again! ah, hope believing,
   It spans the ocean wide,
It sweeps the joy and mirth of living
   Across life's surging tide.

Auf wiedersehen! words so hopeful,
   They give us joy, not pain;
For parting loses half its sorrow
   When friends may meet again.

To meet again! ah, shall we say it,
    When Death, the King, comes nigh?
Or must we look with tears and heart-break,
    And simply say—good-bye?

Oh, death would lose its pangs of sorrow,
    And life its fiercest pain,
If we could stand beside the dying
    And say, *auf wiedersehen!*

THE NEW MOON.

I walked at eve in the twilight dim,
    Where the mists were falling down;
Out to the west from the haunts of men,
    Out of the busy town.
Out of my dream-world, unto the west,
    Looked I with longing eyes
    Up to the deep-stained skies,
With a passion of grief in my throbbing breast.

I saw the sun in a splendor set,
    And the mountains darkly red
With the fires that burned in the boundless west,
    Whither the day had sped.
And the crescent moon in her circle rolled,
    Paling the stars of gold,
    Young — yet, oh, so old,
Measured by time and the ages told,

I marked the glow in the crimson west,
    I marked the pale gold stars,
And the blood-red sun, as he sank to rest,
    Under the western bars.
"Beautiful moon," I cried at last,
    "You, with your heart of gold,
    Young — yet, oh, so old,
Tell me the truths of the ages past!"

My soul is prisoned in dusky rooms,
    And I look from the windows gray
Out of the gloom with longing eyes,
    For the golden light of day.
Silent moon in the twilight sky,
    Give from your boundless store
    Truth—and I ask no more;
Truth from the lives of men gone by.

"The souls of men have uttered their woes
    Unto your heart of gold;
You have heard the plaints of dying and dead,
    And the voices of sages old.
Tell me the truth, that I may know,
    Better than those gone by,
    How men should live and die,
Filled with peace in life's ebb and flow."

The great, mysterious streams of life,
    Pass by my heart's closed door,
And I reach for help as the waves rush on
    Swift to the nevermore.
But vain is my prayer; the stars drop down,
    And the beautiful moon goes by,
    At peace, in the star-crowned sky—
While I—go back to the busy town.

LIFE.

Splash, splash, splash!
      So the brooklets run
To the river wide in the golden sun.

Dash, dash, dash!
      So the rivers wide,
Sweep to the ocean's foaming tide.

Sweet, sweet, sweet!
      So the wood-birds sing,
High in the arch of each forest swing.

Laugh! says the brook in its careless way,
Laugh! to the child who joins in its play.

Weep! moans the river dark and wide,
Weep! to the men who watch by its tide.

Rest! sings the bird with folded wing,
Rest! with the peace that death must bring.

THE MOUNTAIN STAR.

Above the crested peaks one star
   Sails through the dark cloud-sea:
A silent world from worlds apart,
   It yet hath voice for me.
The tall pines whisper to the winds
   The stream is silent, never;
But stern the mountains voiceless stand,
   The stars are mute forever.

I lose the star in pathless fields
   That mark the boundless West,
But not the thought it brings to me,
   Nor yet its dream of rest.
Below the mountain wall it sinks,
   Lost in the orient morning;
But heart, I know, 'twill come again,
   The queen-moon's crown adorning.

If only faith could be as sure
    In all things as this star,
Our lives, with all their loves and joys,
    No thought of death could mar.
But heart, the light of ages, lives
    In that pale star, declining;
Why should not life, then, fading here,
    In other worlds be shining!

SOPRIS PEAK.*

Peak of snow, uplifted
    In the sun-flame's glow,
Eyes that seek thee wonder,
    Hearts grow meek and low.
When at eve thy splendor
    Purpling, fades to dun,
Souls that reach must find thee
    Sphinx-like as the sun.

O'er the ranges, whitened
    By the ermined king;
In the boundless spaces
    Star-worlds silent swing.

---

*One of the highest mountains in Colorado. Its summit is covered with snow throughout the year.

Hast thou not communion —
  Rifted peak, sublime —
With these mystic sky-gods
  In the stretch of time!

Learned thou not the lesson
  Of the ages past:
Suns that burn forever —
  Worlds that could not last?
Has not Nature whispered
  Tales of rock and field,
From the heart of mankind,
  Truths forever sealed!

Peak of snow, majestic,
  Thou art silent, stern,
Voiceless as the ages —
  Yet of thee we learn.

SOPRIS PEAK.

Shining peak, uplifted,
    May not life like thee
Reach to heights undreamed of —
    Pure, and grand, and free!

TWIN BORN.

Two roses hung in the purple eve,
    In the glow of the crimson sun;
And the shadow crept in each golden heart
    When the summer's day was done.

But one was dead when the morning came,
    Down-trodden by careless feet;
In the light of the moon its petals fair
    Had yielded their fragrance sweet.

But the other bloomed in the garden fair,
    Like a maid with a blushing face;
In the midst of the dew-kissed blossoms rare
    It hung with a royal grace.

A lesson of life I learned from these —
    A thought that was bitter-sweet —
As I gazed on the rose that bloomed above
    And the blossom that lay at my feet.

Twin-born are the flowers of hope and love
    As they bloom in the purple eve;
Who knows when the golden day is done
    What shadows the night will weave?

In the crimson heart of the fairest rose
    The shadow of death may lie;
And the hope that lives in the radiant day
    In the gloom of the night may die.

Twin-born are the thoughts and the lives of men,
    Twin-born are the hopes of the heart;
But the morning that comes with its crimson glow
    Sees them faded, and far apart.

THE ELECTRIC LIGHT.

Do you see the flame of that silver light,
    Lifted to heights afar,
Shining at night in the shrouded gloom,
    Proud as a god-like star?
In the lonely street, where the shadows meet,
    Does it bring no loftier thought
    Than the dreams by dreamers wrought

Does it seem to you as it seems to me,
    A symbol of man's great soul,
Reaching to grander, god-given heights,
    As the ages onward roll?
In the star-flames white of its silver-winged light
    Are there written no word for you —
    Old-new thoughts that are true?

THE ELECTRIC LIGHT.

Is there a fate for souls more fair
    Than to give of their own best light,
Flames that are white with a purpose true,
    Flames from the star's still height?
Steady and bright, oh, the far-shining ligh
    Of the soul like the lamp on high,
    Symbol of faith in the sky!

RED MOUNTAIN.*

Oh sacred mount with kingly crest
    Through tideless ether reaching,
The earth-world kneels to hear the prayer
    Thy dusky slopes are teaching.
With mystic glow on sunset eves
All trembling, lie thy blood-red leaves;
Their silken veins with gold inwrought,
Oh, glorious is thy world-wide thought!

A wordless eloquence is thine,
    A spell too deep for speaking;
Thy grand old face to godhood looks
    The gates of heaven seeking.
Thy yellow crown is symbol rare

---

*A mountain in the vicinity of Aspen, Colorado. It is most beautiful in September, when the frost has tinged its slopes with the brilliant tints of autumn.

Of human natures, strong and fair;
And mountain walls that rise and shine
Are Nature's pledge of peace divine!

No darksome storms can dim thy glow,
   The rainbow lights thee then;
In sun or rain thy glorious face
   Is lifted up to men.
From peak to peak the rose-hues sweep
Through azure spaces, infinite, deep;
Oh, sacred mount, augustly fair,
The earth-world kneels to hear thy prayer.

HOPE.

Sweet hope, that shines a radiant star
   Through all our darkest days.
A heavenly light that guides afar
   In narrow, tear-worn ways,
Sweet hope—is there a heart that knows not thee,
Linked with thy sisters, faith and charity?

No artist hand could paint a scene
   Where darkness reigns supreme;
The light—the light must enter in,
   Through darkest night—a gleam:
And thus he paints the glow—the shaft of light
That makes the picture fair—so beauteous to the sight.

No human life is wholly dark—
   God makes not living so;
Against the shade—the shaft of light,
   The bright, eternal glow.
Sweet hope—through all the sorrow and the sin,
The darkness thou dost pierce and enter in!

MY PRAYER.

Oh, not for strength to do some mighty thing
    That human never dared before;
Or giant intellect that thirsts to grasp
    All sciences and worldly lore —
Is this I ask, dear Lord, my plea,
When hushed in prayer, my soul seeks Thee.

Oh, not for fame whose heavy-weighted crown
    Makes weary those who strive to gain,
Nor for the gold that oft a tempter is
    To leave on eager souls its stain —
Is this my prayer, when morn and night,
Mine eyes look upward to the Light.

But simply this—to live each gracious day
    As though it were my last and best:
To do each duty, finish every task —
    In Thy dear hand to leave the rest;
And loving Thee, live in Thy grace,
Contented in the lowliest place.

THE LAKE.

The beautiful lake at noon-day lies
Asleep in the glow of the sun-lit skies;
And fair as a dream, and pure as a gem,
It lies in the mountain's diadem;
Still, oh still, as a heart at rest,
Knowing no pain in the pulseless breast.

At even-tide do the glories brim
Over the still lake's golden rim;
To the dewy shore when the colors sweep,
In a full-tide splendor wide and deep,
And crimson waves do its waves enfold —
Deep, oh, deep, in its heart of gold.

But, fairer still is the lake at night,
In the dream-like spell of the soft moonlight:
When the stars encircle the mountains old
With their jeweled crown of shining gold,

And the shadows drift in its close embrace,
Low on the shore in each vine-clad place.

Oh, the dreamy spell of the dreamy night,
When the boats float down in the silver light,
And the oar-blades flash and keep slow time
To the music's dreamy reach and rhyme,
And the soul outpours in a burst of song —
As the moon in the water drifts along.

Oh, the dip of the oar and the music's swell —
How sweet, how sweet, is their dream-like spell!
And the ripple of moon-light in the wake
Of the bird-like boat that skims the lake.
Oh, these are the sweets in the bitter of life
That calms its fever and stills its strife.

For every joy there are deeps of pain,
And pleasures once gone come ne'er again;
But this I know, that in memory's hall,
Some pictures fade not beyond recall;
And I long to feel that these dreams will last,
Though in sterner scenes our lives are cast.

COMING.

There is a dream of budding blooms
    And scent of roses sweet;
Soft daffodils that swift unfold
    Beneath Spring's light-winged feet.

All life-blessed things rejoice and smile—
    Dear Nature builds anew—
And know ye not, oh, care-worn hearts,
    Life, too, hath hope for you?

THE RANCHMAN.

" And I have said, and I say it ever,
  As the years go on and the world goes over,
  'Twere better to be content and clever
    In the tending of cattle and tossing of clover,
    In the grazing of cattle and growing of grain,
    Than a strong man striving for fame or gain."
                              —*Joaquin Miller*.

There's nothing so sweet in the world as this,
To wake in the morn with the sun's first kiss
    On the dewy clover;
When the grass is wet and the flaming rose
Lifts her lovely head where the reaper sows
    The wide fields over.

The blossoming breath of the upturned sod
Is the weetest incense offered to God
    From our mother-earth.
And lo! from the tree-tops rising higher,
Are the musical strains of the bright bird-choir
    In the dawn's swift birth.

Where the rushing streams like spirits leap
Down the mountain cañons dark and steep
    Is joy forever;
And a God-given life is this to live:
Away from the toil that the cities give,
    And their vain endeavor.

In the heart of the wilds where nature's wealth
Brings the guerdon of joy and bounding health—
    Is found true pleasure.
To follow the plow and sheave the grain,
In this the spirit finds real gain
    And the heart full measure.

Oh, sun-browned men with your hearts of fire,
And your strong brave arms that do not tire,
    To you is given
The highest joy and the purest mirth—
The noblest gifts to be gained on earth
    This side of heaven.

THE LADDER OF PRAYER.

It reaches to God, the ladder of prayer!
    Worn are its steps and old,
For countless the climbers who toil up there—
    Numbers and numbers untold.
Yet still it is fair in the sight of God,
Leading to heaven from earth-worn sod.
      Each step is an altar
      Where weak souls falter,
To sob out a heart-broken prayer.

Sometimes, in the shadow, the ladder of prayer
    On the windows of heaven doth lean,
And the souls who have climbed to the spirit's height
    Catch gleams of a glorious sheen.
For it leads to the pure, the true and bright,
To all things heavenly, grand and right.
      Would we rise from the sod,
      We must climb to God
On the beautiful ladder of prayer!

It is wet with tears, the ladder of prayer,
  But each tear is a priceless gem ;
We shall wear them all, dear hearts, some day,
  In a heavenly diadem.
With the weight of pain each step is worn,
And sad are the souls who climb and mourn.
  But all gifts that are rare
  We will find up there,
On the beautiful ladder of prayer!

In the darkest night, on the ladder of prayer
  We may reach to God and Light,
And the humblest toiler may jostle the king
  And each be hidden from sight.
But all may mount who would seek to climb :
God hears each prayer in His own good time.
  And His angels fair
  Help the toilers there —
Up the beautiful ladder of prayer!

www.ingramcontent.com/pod-product-compliance
Lightning Source LLC
Chambersburg PA
CBHW020257170426
43202CB00008B/412